**Ankita Goyal
Bhuvneshwer Suthar**

Semi Empirical Mass Formula

A Review

Anchor Academic
Publishing

Goyal, Ankita, Suthar, Bhuvneshwer: Semi Empirical Mass Formula. A Review,
Hamburg, Anchor Academic Publishing 2016

Buch-ISBN: 978-3-96067-001-8
PDF-eBook-ISBN: 978-3-96067-501-3
Druck/Herstellung: Anchor Academic Publishing, Hamburg, 2016

Bibliografische Information der Deutschen Nationalbibliothek:
Die Deutsche Nationalbibliothek verzeichnet diese Publikation in der Deutschen
Nationalbibliografie; detaillierte bibliografische Daten sind im Internet über
http://dnb.d-nb.de abrufbar.

Bibliographical Information of the German National Library:
The German National Library lists this publication in the German National Bibliography.
Detailed bibliographic data can be found at: http://dnb.d-nb.de

All rights reserved. This publication may not be reproduced, stored in a retrieval system
or transmitted, in any form or by any means, electronic, mechanical, photocopying,
recording or otherwise, without the prior permission of the publishers.

Das Werk einschließlich aller seiner Teile ist urheberrechtlich geschützt. Jede Verwertung
außerhalb der Grenzen des Urheberrechtsgesetzes ist ohne Zustimmung des Verlages
unzulässig und strafbar. Dies gilt insbesondere für Vervielfältigungen, Übersetzungen,
Mikroverfilmungen und die Einspeicherung und Bearbeitung in elektronischen Systemen.

Die Wiedergabe von Gebrauchsnamen, Handelsnamen, Warenbezeichnungen usw. in
diesem Werk berechtigt auch ohne besondere Kennzeichnung nicht zu der Annahme,
dass solche Namen im Sinne der Warenzeichen- und Markenschutz-Gesetzgebung als frei
zu betrachten wären und daher von jedermann benutzt werden dürften.

Die Informationen in diesem Werk wurden mit Sorgfalt erarbeitet. Dennoch können
Fehler nicht vollständig ausgeschlossen werden und die Diplomica Verlag GmbH, die
Autoren oder Übersetzer übernehmen keine juristische Verantwortung oder irgendeine
Haftung für evtl. verbliebene fehlerhafte Angaben und deren Folgen.

Alle Rechte vorbehalten

© Anchor Academic Publishing, Imprint der Diplomica Verlag GmbH
Hermannstal 119k, 22119 Hamburg
http://www.diplomica-verlag.de, Hamburg 2016
Printed in Germany

PREFACE

This can be very fruitful book to the students and research scholar of the field of nuclear physics as well as physics to understand how to start the research work. Here a very common topic 'Semi empirical Mass Formula' of nuclear physics is chosen to start work to research. The experience was very nice during this work to understand the problems and the way to solve them. One of author (Ankita) express her gratitude to all the people at Govt. Engineering College, Bikaner who inspire of their busy schedule took personal interest to ensure that the project of mine was a through learning process for me. I am delighted with the project and the experience gained by me during this period.

We would also like to thank all my friends who have helped us throughout the project. We would like to also greatful to blessing of our parents and family members. One of author (BS) wishes to place on record his profound gratitude to my wife Mrs. Shalini Suthar for their moral support. Last but not list the loving from my little angels Jyotasna and Khushi is make me fresh and relax.

Ankita

Bhuvneshwer Suthar

List of Paper Presented

Ankita and B. Suthar, *"A Review on Semi Empirical Mass Formula"*, presented in National Conference on 'Recent Advances in Materials Science and Technology', at Govt. MLV College, Bhilwara during Dec. 22-23, 2014.

Table of Contents

Preface	i
List of Paper Presented	ii
Chapter 1 Introduction	1
1.1 Introduction	1
1.2 Need of Modeling	2
1.3 Classification of Nuclear Models	3
1.3.1 An individual particle model	3
1.3.2 Strong interaction model	4
1.4 Historical Development	5
1.5 Outline of Project	8
Chapter 2 Semi Empirical Mass Formula	10
2.1 Liquid drop model	10
2.2 Semi Empirical Mass Formula	11
2.2.1 Volume energy	12
2.2.2 Surface energy	14
2.2.3 Coulomb Energy	15
2.2.4 Asymmetry energy	17
2.2.5 Pairing energy	19
Chapter 3 Comparative Study of Binding Energy	22
3.1 Introduction	22
3.2 Binding Energy per nucleon	23

3.3	Total Binding Energy	25
3.4	Comparative Study of Binding Energy	26
Chapter 4	Conclusions	33
Bibliography		35
About the Authors		38

Chapter 1

Introduction

1.1 Introduction

Nuclear physics is the field of physics that studies the constituents and interactions of atomic nuclei. The most commonly known applications of nuclear physics are nuclear power generation but the research has provided application in many fields, including those in nuclear medicine and magnetic resonance imaging, nuclear weapons, ion implantation in materials engineering, and radiocarbon in geology and archaeology.

The field of particle physics evolved out of nuclear physics and is typically taught in close association with nuclear physics.

A heavy nucleus can contain hundreds of nucleons which mean that with some approximation it can be treated as a classical system, rather than a quantum-mechanical one. In the resulting liquid-drop model, the nucleus has an energy which arises partly from surface tension and partly from electrical repulsion of the protons. The liquid-drop model is able to reproduce many features of nuclei, including the general trend of binding energy with respect to mass number, as well as the phenomenon of nuclear fission.

Superimposed on this classical picture, however, are quantum-mechanical effects, which can be described using the nuclear shell model, developed in large part by Maria Goeppert-Mayer and J. Hans D. Jensen. Nuclei with certain numbers of neutrons and protons (the magic numbers 2, 8, 20, 28, 50, 82, 126, ...) are particularly stable, because their shells are filled.

Other more complicated models for the nucleus have also been proposed, such as the interacting boson model, in which pairs of neutrons and protons interact as bosons, analogously to Cooper pairs of electrons.

1.2 Need of Modeling

During the study of nuclear two body problem and inter nucleon forces it has been emphasized that there exists no derivation of nucleon-nucleon force from first principles. From this study however we have been able to obtain considerable information about the nature of this force. This force has turned out to be extremely complicated and very ill behaved from a numerical point of view showing strong repulsion or hard core at short distance between nucleons. The nucleons in nucleus do not feel the bar nucleons–nucleons and there seem to be many body forces.

The term nuclear structure includes information about all the properties of the nucleus such as their spin, parity, electric and magnetic moments, charge radius, excited states etc. Theoretical determination of all this information requires a calculation of the total wave function of the nucleus. For nuclear systems with A>2 one is faced with a number of difficulties in this endeavor. The first difficulty has already been emphasized that we do not as yet born a clear conception of the nature of nuclear interaction. The second difficulty is that even if we assume forces of simple form though containing aspect of its character such as short range, spin dependence, tensor components, etc. There is at present no simple approximation procedure using which one can obtain reasonably accurate solution of many- body Schrodinger equation. This realization has driven the investigators from early days of nuclear physics to look for conceptual or mathematical models based on certain simplifying assumption which without too heavy and involved calculation could provides more or less satisfactory description of various characteristic of nuclear structure and nuclear properties.

1.3 Classification of Nuclear Models

Nuclear models are classified in two types which are as follow

1.3.1 An individual particle model with nucleons in discrete energy states in which the behavior of individual particles within a nucleus determines the characteristic of the nucleus as a whole and the nucleons move nearly independently in a common nuclear potential.

i. Potential well model

ii. The optical model

iii. Fermi gas model

iv. Shell model

v. Single particle shell model

vi. Many particle shell model

vii. Individual particle model

viii. J-j coupling model

ix. Unified model

x. Collective model

xi. Rotational model

xii. Spheroid –core model

xiii. Super fluid model

1.3.2 Strong interaction model, in which the constitute nucleons are taken to be strongly coupled to one another because of their strong and short range interaction. A collective model with no individual particle states: An example is the Liquid Drop Model which is the basis of the semi-empirical mass formula.

i. Compound nucleus model
ii. Liquid drop model
iii. Alpha-particle model
iv. Cluster model

1.4 Historical Development

For the study to explain the properties of nuclei and their structure and internal motion, one has to require modeling, in nuclear physics. There are several conceptual or mathematical models based on simple assumption without too heavy calculation, could provide more or less satisfactory description of various characteristic of nuclear structure and nuclear property. One of them is Liquid Drop Model purposed by George Germow in 1928. In nuclear physics, the semi empirical mass formula is used to find nuclear masses, binding energy and other properties of nuclei which is based on liquid drop model and firstly formulated by C. F. V. Wcizsackcr and after then modified by Bethe and some other physician. It is based partially on the theory and partially on empirical measurement so this formula is called semi empirical mass formula.

The Semi Empirical Mass Formula (SEMF) plays a significant role in the development of nuclear physics and it involved five kind of energy namely, the volume, surface, coulomb, asymmetry and pairing energy corresponding five type of coefficient, although refinements have been made to the coefficients over the years, the structure of the formula remains the same today. Over the past few decades, SEMF has sometimes been extended by adding extra terms or sometimes been modified by changing the dependence of each term on A and Z slightly in an effort to predict the nuclear binding energies as accurately as possible.

In 1958 D. William et. al. modifies Weizsacker's SEMF to include effects of the diffuse nuclear surface indicated by recent electron scattering experiments. Volume and surface effects are combined by integrating over an assumed trapezoidal density function similar to that found experimentally and get good fits with experimental data. In 1962 R. Ayres et. al. gives a new SEMF for binding energy of nuclear species resembling the usual bathe Weizsacker formula. In 1965 M. Bauer et. al. show the suitable rearrangement for the ground state energy of a super fluid nucleus can reproduce SEMF. They also concluded that the coefficients are not constant but depend on the deformation, and being larger for deformed nuclei. In 1965 N. Zeldes et. al. revised the mass table (1961) using least squares adjustment including newer experimental data, then difference of masses were plotted as a function of N or Z and formulated a generalized shell model mass formula.

In 1966 D. William et. al. represented a semi empirical mass theory to consider the potential energy as a function of N, Z and the nuclear shape with assumption given by the liquid drop model, modified by shell correction as the resultant formula has 7 adjustable parameter in which 4 due to LDM part and 3 due to shell correction. In 1971 H.A. Mavromatis calculated the coefficients of the volume, surface and symmetry terms in the SEMF from a first-order shell-model calculation using the Sussex matrix elements. In 1976 D. William et. al. developed the droplet model and formulated new SEMF with 16 coefficients, based on the literature and simple physical arguments. Some of these coefficient determined by fitting of masses, deformation etc.

In 1980 C.Y. Tseng et. al. presented a new formula which has forms for the Coulomb-energy, symmetry-energy and pairing-energy terms different from the conventional ones. It gives good agreement with experiment for the β-stability position and calculated binding energy. In 1986 X. Fuxin presented a new binding energy formula using 1271 nuclei masses exactly measured by the least square method; determine the parameters in order to obtain the concrete form of the new formula.

In 2002 C. Samanta **et. al.** extended the Bethe-Wisecracker (BW) mass formula to light nuclei and some new shell closures have been identified. This modified BW mass formula explains the shapes of the binding energy versus neutron number curves of most of the elements from Li to Bi. In 2003 D.N. Basu concluded that the extension of Bethe-Weizsacker mass formula to light nuclei is beset with some difficulties. There are predictions/ about the stability of some light nuclei which are not in agreement with the experimental observations. In 2004 Chris Daley determined the optimum values of the coefficients that appear in SEMF using a genetic algorithm (GA) and investigated mass formula which includes an additional term that provides a surface asymmetry correction.

In 2008 Michael W. Kirson extended SEMF to include a number of additional terms with all possible combinations of extra terms were fitted in turn to the measured nuclear masses and the results analyzed in order to reveal correlations and mutual influences between the various terms. In 2010 Byeongnoh Kim et. al. examined whether the mass number A, dependence of each of the five terms in SEMF would still be valid for nuclei very far from the stability line in the chart of nuclei.

1.5 Outline of Project

In this project, we have discussed the review of semi empirical mass formula and the comparative study of nuclear binding energy. In the chapter 2, we will discuss the liquid drop model and semi empirical mass formula based on this model. The various energy terms also will be presented there. In the chapter 3, the various energy terms per nucleon with mass number will be presented graphically and discussed. The comparative study of binding energy per nucleon theoretically calculated and experimental data will be presented and discussed.

Chapter 2

Semi Empirical Mass Formula

2.1 Liquid drop model

The liquid drop model in nuclear physics treats the nucleus as a drop of incompressible nuclear fluid. It was first proposed by George Gamow and then developed by Niles Bohr and John Archibald Wheeler. The fluid is made of nucleons (protons and neutrons), which are held together by the strong nuclear force. This is a crude model that does not explain all the properties of the nucleus, but does explain the spherical shape of most nuclei. It also helps to predict the nuclear binding energy. The essential assumptions are;

1) The nucleus consists of incompressible matter.

2) The nuclear force is identical for every nucleon.

The nuclear force saturates, thus one might inquire whether a nucleus can be represented as a crystalline aggregate of nucleons.

The liquid drop model, developed from the observation of similar properties between a nucleus and a drop of incompressible fluid, helps explain nuclear phenomena such as the energetic of nuclear fission and the binding energy of nuclear ground levels which cannot be illustrated by the shell model. In view of similarities such as the latent heat of vaporization of fluid which is comparable to the constant binding energy per nucleon, and the surface tension effects of nucleus as well as a liquid drop, the quantitative aspect of the model delivers a formula that approximates the mass and binding energy of nuclei.

More specifically, heat of vaporization represents the amount of energy required to convert molecules from liquid phase to gas phase. The latent heat of vaporization is proportional to the number of molecules in the liquid. The binding energy of nucleus displays a similar relationship where it is proportional to the number of nucleons.

2.2 Semi Empirical Mass Formula

Mathematical analysis of the liquid drop model delivers an equation which attempts to predict the binding energy of a nucleus in terms of the numbers of protons and neutrons it contains. This equation has five terms on its right hand side. These correspond to the cohesive binding of all the nucleons by the strong nuclear force, the electrostatic mutual repulsion of the protons, a surface energy term, an asymmetry term (derivable from the protons and neutrons occupying independent quantum momentum states) and a pairing term (partly derivable from the protons and neutrons occupying independent quantum spin states).

In the following formulae, let A be the total number of nucleons, Z the number of protons, and N the number of neutrons.

The mass of an atomic nucleus is given by

$$m = Zm_p + Nm_n - \frac{E_B}{c^2} \qquad (2.1)$$

Where m_p and m_n are the rest mass of a proton and a neutron, respectively, and E_B is the binding energy of the nucleus.

The semi-empirical mass formula states that the binding energy will take the following form:

$$E_B = a_V A - a_S A^{2/3} - a_C \frac{Z^2}{A^{1/3}} - a_A \frac{(A-2Z)^2}{A} - \delta(A,Z) \qquad (2.2)$$

Each of the terms in this formula has a theoretical basis, as will be explained below.

2.2.1 Volume energy

When an assembly of nucleons of the same size is packed together into the smallest volume, each interior nucleon has a certain number of other nucleons in contact with it. So, this nuclear energy is proportional to the volume.

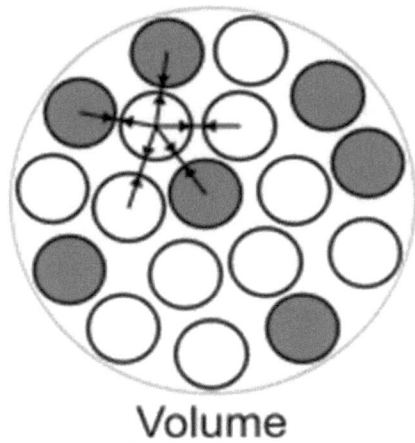

Volume

Figure 2.1 Volume Energy

Volume energy term defined as

$$E_v = a_V A \qquad (2.3)$$

The volume of the nucleus is proportional to A, so this term is proportional to the volume, hence the name. The basis for this term is the strong nuclear force. The strong force affects both protons and neutrons, and as expected, this term is independent of Z. Because the number of pairs that can be taken from A particles is proportional to A^2. However, the strong force has a very limited range, and a given nucleon may only interact strongly with its nearest neighbors and next nearest neighbors. Therefore, the number of pairs of particles that actually interact is roughly proportional to A, giving the volume term its form.

The coefficient a_V is smaller than the binding energy of the nucleons to their neighbors E_b, which is of order of 40 MeV. This is because the larger the number of nucleons in the nucleus, the larger their kinetic energy is, due to the Pauli Exclusion Principle.

2.2.2 Surface energy.

A nucleon at the surface of a nucleus interacts with fewer other nucleons than one in the interior of the nucleus and hence its binding energy is less. This surface energy term takes that into account and is therefore negative and is proportional to the surface area.

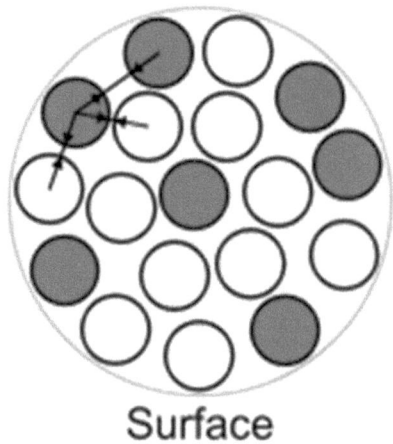

Figure 2.2 Surface Energy

The surface energy term is defined as

$$Es = a_S A^{2/3} \quad (2.4)$$

This term, also based on the strong force, is a correction to the volume term.

The volume term suggests that each nucleon interacts with a constant number of nucleons, independent of A. While this is very nearly true for nucleons deep within the nucleus, those nucleons on the surface of the nucleus have fewer nearest neighbors, justifying this correction.

This can also be thought of as a surface tension term, and indeed a similar mechanism creates surface tension in liquids.

If the volume of the nucleus is proportional to A, then the radius should be proportional to $A^{1/3}$ and the surface area to $A^{2/3}$. This explains why the surface term is proportional to $A^{2/3}$. It can also be deduced that a_S should have a similar order of magnitude as a_V.

2.2.3 Coulomb Energy.

The electric repulsion between each pair of protons in a nucleus contributes toward decreasing its binding energy.

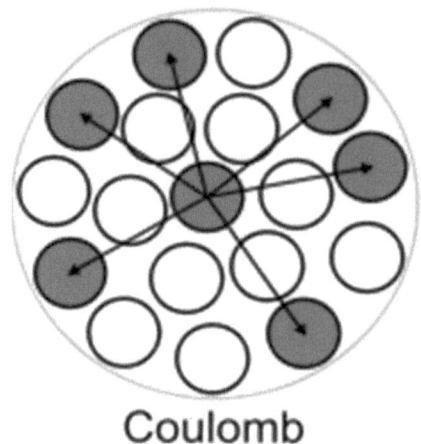

Coulomb

Figure 2.3 Coulomb Energy

The coulomb energy term is defined as

$$\text{Ec} = a_C \frac{Z^2}{A^{1/3}} \tag{2.5}$$

The basis for this term is the electrostatic repulsion between protons. To a very rough approximation, the nucleus can be considered a sphere of uniform charge density.

The potential energy of such a charge distribution can be shown to be

$$E = \frac{3}{5}\left(\frac{1}{4\pi\epsilon_0}\right)\frac{Q^2}{R} \tag{2.6}$$

Where Q is the total charge and R is the radius of the sphere. Identifying Q with Ze, and noting as above that the radius is proportional to $A^{1/3}$, we get close to the form of the Coulomb term. However, because electrostatic repulsion will only exist for more than one proton, Z^2 becomes $Z(Z-1)$. The value of a_C can be approximately calculated using the equation above.

2.2.4 Asymmetry energy (also called Pauli Energy)

Energy associated with the Pauli Exclusion Principle. Were it not for the Coulomb energy, the most stable form of nuclear matter would have the same number of neutrons as protons, since unequal numbers of neutrons and protons imply filling higher energy levels for one type of particle, while leaving lower energy levels vacant for the other type.

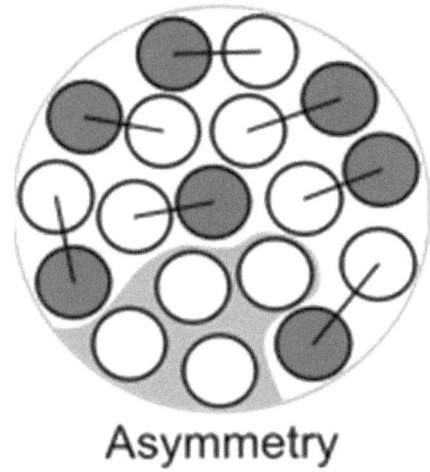

Figure 2.4 Asymmetry Energy

The asymmetry energy term is defined as

$$Ea = a_A \frac{(A-2Z)^2}{A} \qquad (2.7)$$

Note that as $A = N + Z$, the parenthesized expression can be rewritten as $(N - Z)$. The form $(A - 2Z)$ is used to keep the dependence on A explicit, as will be important for a number of uses of the formula.

The theoretical justification for this term is more complex. The Pauli Exclusion Principle states that no two fermions can occupy exactly the same quantum state in an atom. At a given energy level, there are only finitely many quantum states available for particles. What this means in the nucleus is that as more particles are "added", these particles must occupy higher energy levels, increasing the total energy of the nucleus (and decreasing the binding energy). Note that this effect is not based on any of the fundamental forces (gravitational, electromagnetic, etc.), only the Pauli Exclusion Principle.

Protons and neutrons, being distinct types of particles, occupy different quantum states. One can think of two different "pools" of states, one for protons and one for neutrons. Now, for example, if there are significantly more neutrons than protons in a nucleus, some of the neutrons will be higher in energy than the available states in the proton pool. If we could move some particles from the neutron pool to the proton pool, in other words change some neutrons into protons, we would significantly decrease the energy. The imbalance between the number of protons and neutrons causes the energy to be higher than it needs to be, for a given number of nucleons. This is the basis for the asymmetry term. One can also understand the asymmetry term intuitively, as follows. It should be dependent on the absolute difference $|N - Z|$, and the form $(A - 2Z)^2$ is simple and differentiable, which is important for certain applications of the formula. In addition, small differences between Z and N do not have a high energy cost. The A in the denominator reflects the fact that a given difference $|N - Z|$ is less significant for larger values of A.

2.2.5 Pairing energy.

An energy which a correction term that arises from the tendency of proton pairs and neutron pairs to occur. An even number of particles is more stable than an odd number.

Pairing

Figure 2.5 Pairing Energy

The term $\delta(A, Z)$ is known as the *pairing term* (possibly also known as the pair wise interaction). This term captures the effect of spin-coupling. It is given by:[3]

$$\delta(A, Z) = \begin{cases} -\delta_0 & Z, N \text{ even } (A \text{ even}) \\ 0 & A \text{ odd} \\ +\delta_0 & Z, N \text{ odd } (A \text{ even}) \end{cases} \quad (2.8)$$

Where

$$\delta_0 = \frac{a_P}{A^{1/2}} \quad (2.9)$$

Due to Pauli Exclusion Principle the nucleus would have a lower energy if the number of protons with spin up were equal to the number of protons with spin down. This is also true for neutrons. Only if both Z and N are even can both protons and neutrons have equal numbers of spin up and spin down particles. This is a similar effect to the asymmetry term.

The factor $A^{-1/2}$ is not easily explained theoretically. The Fermi ball calculation we have used above, based on the liquid drop model but neglecting interactions, will give A^{-1} dependence, as in the asymmetry term. This means that the actual effect for large nuclei will be larger than expected by that model. This should be explained by the interactions between nucleons.

Chapter 3

Comparative Study of Binding Energy

3.1 Introduction

Nuclear binding energy is the energy required to split the nucleus of an atom into its component parts. The component parts are neutrons and protons, which are collectively called nucleons. The binding energy of nuclei is usually a positive number, since most nuclei require net energy to separate them into individual protons and neutrons. Thus, the mass of an atom's nucleus is usually less than the sum of the individual masses of the constituent protons and neutrons when separated. This notable difference is a measure of the nuclear binding energy, which is a result of forces that hold the nucleus together. During the splitting of the nucleus, some of the mass of the nucleus gets converted into huge amounts of energy (according to Einstein's equation $E=mc^2$) and thus this mass is removed from the total mass of the original particles, and the mass is missing in the resulting nucleus. This missing mass is known as the mass defect, and represents the energy released when the nucleus is formed.

In this chapter, we will discuss the calculated nuclear binding energy using the semi empirical mass formula, which has been discussed in detail in last chapter. Here, the calculated nuclear binding energy is compared with the experimental data, which shows very good agreement.

3.2 Binding Energy per nucleon

In the last chapter, we have explained the semi empirical mass formula in detail, which is based on liquid drop model. These contain varies energy terms like volume energy, surface energy, coulomb energy, asymmetric energy and pairing energy. For study the effect of these energies on the binding energy we plot binding energy with mass number by using calculated values. The graph between the various energy terms per nucleon with mass number is plotted using equation (2.3-2.7), which is shown in figure 3.1.

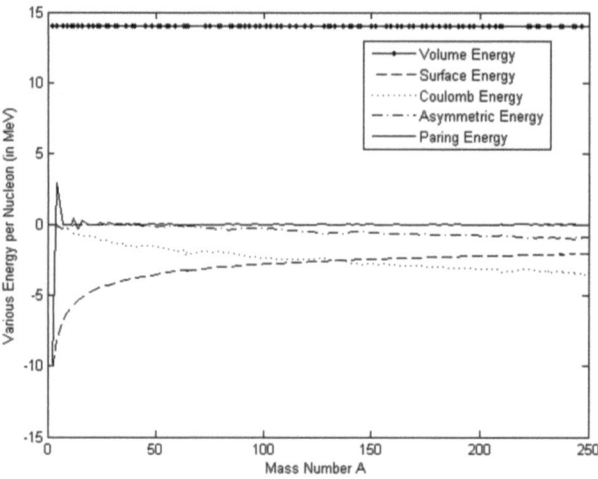

Figure 3.1. The graph between the varies energy terms (per nucleon) with mass number.

The above figure shows clearly seen that the volume energy term per nucleon is constant with mass number A and have positive values. While the rest of all terms contribute negative values in nuclear binding energy as shown in above figure. The surface energy values have larger values for light weight nuclei and decreases with mass number A. The coulomb energy increases with mass number A and get significant values for high weight nuclei, which decreases the total nuclear binding energy values in higher weight region. Due to this, the stability of higher weight nuclei decreases with mass number. The asymmetric energy varies with mass number and attains highest values around 1Mev while the pairing energy has very less contribution in certain nuclei.

3.3 Total Binding Energy

In this section, we have plotted a graph between the experimental values of binding energy per nucleon and mass number A, which is shown in figure 3.2(a).These experimental data have been taken from 1995 update to the atomic mass evaluation. We calculate the values of binding energy per nucleon using SEMF equation (2.2) for stable nuclei chosen according to periodic table. For this purpose, various coefficients of semi empirical mass formula are taken from table 1. The plot between the calculated values of the binding energy per nucleon and mass number A is plotted in Figure 3.2(b). Here the pairing energy term is not included in total binding energy calculation.

Table: 1 various energy coefficients values of semi empirical mass formula.

S. No.	a_v	a_s	a_c	a_a	
1.	15.3	16.7	0.69	22.6	1955 R.D. Evans
2.	14.66	15.4	0.602	20.5	1942 Fluegee
3.	13.86	13.2	0.58	19.5	1936 Bethe and Bacher
4	15.75	17.8	0.711	23.7	1995 Rohlf

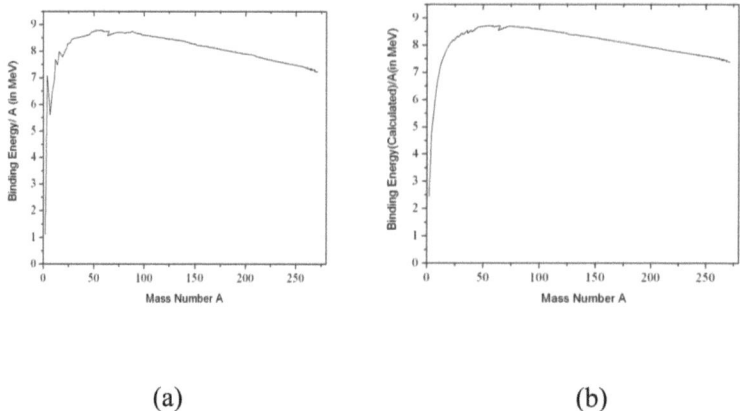

(a)   (b)

Figure 3.2 The plot between Binding Energy per nucleon in Mass Number A for (a) experimental data and (b) calculated values using SEMF.

From the above graphs, it is clearly seen that the experimental and calculated values of the binding energy per nucleon are very much identical. The semi-empirical mass formula provides a good agreement with heavier nuclei, while poor with very light nuclei. This is because the formula does not consider the internal shell structure of the nucleus. For light nuclei, it is usually better to use a model that takes this structure into account.

3.4 Comparative Study of Binding Energy

In this section, we measured the binding energy per nucleon for more stable nuclei chosen according to the periodic table (of 110 elements) using semi empirical mass formula of bathe - Weizacker with different coefficients given by various researcher then calculated binding energy per nucleon is compared with the experimental binding energy per nucleon obtained from 1995 update to the atomic mass evaluation. And error curves are plotted as shown below.

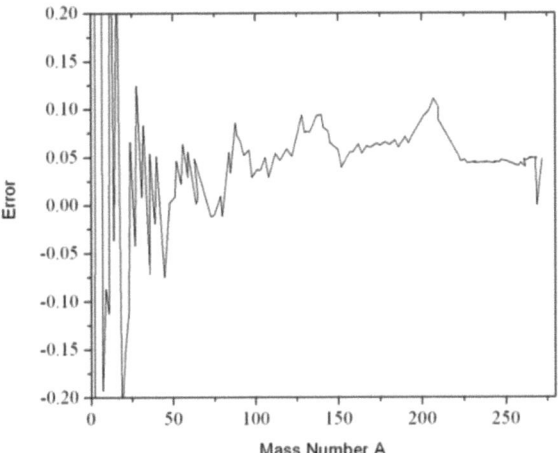

Figure 3.3 The plot between Error in Binding energy with mass number (for coefficients given by R.D. Evans 1955).

The above graph is plotted for the coefficients given by R.D. Evans in 1955. We observed from the above graph that the difference between experimental data and calculated values of binding energy per nucleon (error) with these coefficients for the stable nuclei varies from -0.93291 to 2.511595 and the average error is 0.056445. For the light weight nuclei of mass number 2 to 45 the error is very high and varies from 2.511595 to -0.93291 and the average error is 0.080408.

For the mid weight nuclei the error is still high and vary with mass number A but become low compare than previously .For a range of mass number 48 to 150 the error varies from -0.075001 to 0.094709 and the average error is 0.0433868.

For the heavy weight nuclei the error is still high but almost constant with the mass number and low compare than previously.

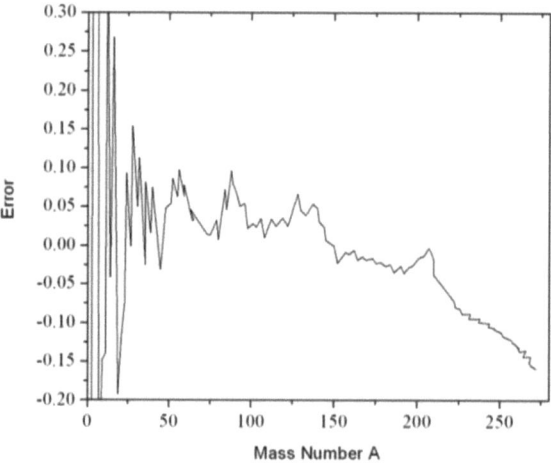

Figure 3.4 The plot between Error in Binding energy with mass number (for coefficients given by Fluegee 1942).

This graph is plotted for the coefficients given by Fluegee in 1942.

From the above graph, it is clearly seen that the region below the mass number 31 shows the very much deviated values of binding energy form the experimental data. So, the semi empirical mass formula with these constants is not useful to study in the region of light weight nuclei.

Here, it was also been observed for the light weight nuclei the error is high, vary from -1.32473 to 2.304928 and the average error is 0.061922. But for mid weight and heavy weight nuclei is shown good agreement with experimental data. As for mass number from 32 to 145 the error for the binding energy varies from -0.03109 to 0.113064 and the average error is 0.042456, while for the mass number from 145 to 272 errors vary from -0.15889 to -0.00319 and average error is -0.06977.

But for a short range of mass number 145 to 195 error becomes very low nearly about zero. The average error is 0.048615. We have obtained the binding energy per nucleon values for a whole range of nuclei are as error varies from -1.32473 to 2.305873 for all most stable nuclei and the average error is -0.00551.

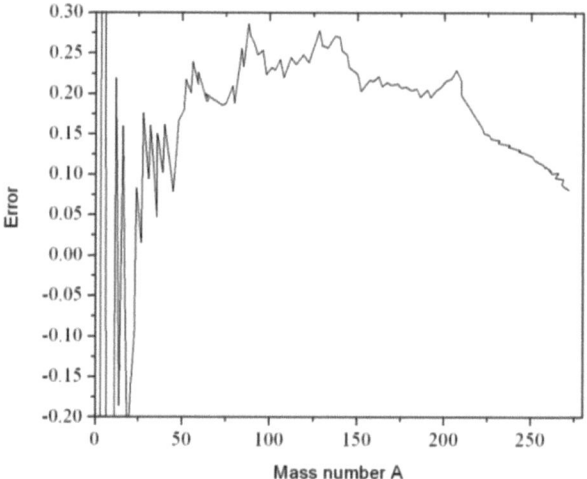

Figure 3.5 The plot between Error in Binding energy with mass number (for coefficients given by Bethe and Bacher 1936).

The above graph is plotted for the coefficients given by Bethe and Bacher in 1936. As shown in above graph for the light weight nuclei these coefficients are not useful. The error is high everywhere but the error can be minimize for mid weight and heavy weight nuclei by changing the volume coefficient.

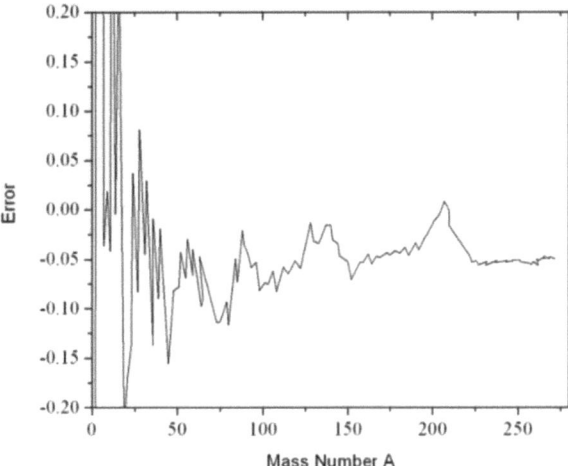

Figure 3.6 The plot between Error in Binding energy with mass number (for coefficients given by Rohlf 1995).

The above graph is plotted for the coefficient given by Rohlf in 1995. It is clear that from the above graph when we measured the binding energy per nucleon using the semi empirical mass formula with these coefficients for all the stable nuclei then we observed the error varies from -0.50984 to 2.761167and the average error is -0.02207.

For the light weight nuclei of mass number 2 to 48 error is very high and varies from -0.50984 to 2.761167. In this region average error is 0.088312. But a range of mass number from 51 to 153 the error become low compare than previously but still varies with the mass number. In this region error varies from -0.01307 to -0.11652 and the average error is -0.05782.

For the heavy weight nuclei of mass number 159 to 272 the error become low in this region the varies from 0.008317 to -0.05615 and the average error is -0.02207. Here, a short range of mass number 210 to 272 the error become almost constant and nearly about -0.05.

Chapter 4

Conclusions

In this project, we have presented a need of nuclear modeling and a review on semi empirical mass formula, which is based on liquid drop model. This formula has been used to calculate the nuclear binding energy with mass number A. The conclusions of the present work are as follow

- The various energy terms of semi empirical mass formula has been calculated and plotted with mass number A. This shows the coulomb energy terms get significant values in higher weight nuclei region and due to their negative contribution higher weight nuclei becomes unstable due to decrease in total binding energy.
- The nuclear binding energy per nucleon calculated values shows very good agreements with experimental values. For the comparative study, the nuclear binding energy values were calculated for various constants given by different researchers.
- The comparative study between the calculated and experimental values shows that none of these coefficients are useful directly for

light weight nuclei because shows much deviation from the experimental data.

- The coefficients given by Bethe et. al. shows much deviation from experimental data for a whole range of elements chosen while the rest of all were in good agreement with experimental data for mid weight nuclei.

- Whereas the coefficients given by Fowler et al. and Rohlf, J. W. et al. show good agreement with experimental values for both mid weight and heavy weight nuclei range, which can be used for study the nuclear properties in this range among these coefficients.

Bibliography

1. Von Weizsacker, C. F. (1935). "Zur Theorie der Kernmassen". Zeitschrift für Physik (in German) 96 (7–8): 431458. Bibcode: 1935ZPhy...96...431W. doi10.1007/BF01337700

2. Bailey, D. "Semi-empirical Nuclear Mass Formula". PHY357: Strings & Binding Energy. University of Toronto. Retrieved 2011-03-31

3. WILLIAM D. GUNTER, JR., AND ROBERT A. HUBBS (1958) "Revised Weizsacker Semi empirical Formula for Diffuse Nuclear Surfaces" Stanford University, Stanford, California,, VOLUME 113, NUMBER 1

4. R. Ayres, W. F. Hornyak, L. Chan, H. Fann "A new semi-empirical mass formula " 01/1962; 29:212-240.

5. M. Bauer, V. Canuto "The semi-empirical mass formula and the superfluid model of nuclei" . 01/1965; 72(1):33-48

6. N. Zeldes, M. Gronau, A. Lev "Shell-model semi-empirical nuclear masses (I) Volume 63, Issue 1, March 1965, Pages 1–75

7. William D. Myers, Wladyslaw J. Swiatecki Nuclear masses and deformations Volume 81, Issue 1, June 1966, Pages 1–60

8. H.A. Mavromatis† The volume, surface and symmetry terms in the semi-empirical mass formula Volume 162, Issue 3, 22 February 1971, Pages 648–656, doi:10.1016/0375-9474(71)90262-4

9. William D. Myers Development of the semi empirical droplet model doi:10.1016/0092-640X(76)90030-9 Volume 17, Issues 5–6, May–June 1976, Pages 411–417

10. C.Y. Tseng, T.S. Cheng F.C. Yang doi:10.1016/0375-9474(80)90611-9 Further study on nuclear mass formulas Volume 334, Issue 3, 11 February 1980, Pages 470–476

11. Xu Fuxin, International Centre for Theoretical Physics, Trieste, Italy, International Atomic Energy Agency and United Nations Educational Scientific and Cultural Organization, INTERNATIONAL CENTRE FOR THEORETICAL PHYSICS, IC/86/52 INTERNAL REPORT

12. C. Samanta and S. Adhikari, Phys. Rev. C 65, 037301 (2002)

13. D.N. Basu, Variable Energy Cyclotron Centre, 1/AF Bidhan Nagar, Kolkata 700 064, India, On the extension of the Bethe-Weizsacker mass formula to light nuclei, September 23, 2003)

14. Chris Daley ,URN: 1049453 An improved Mass formula, Department of Physics The University of Surrey

15. Michael W. Kirson, Mutual influence of terms in a semi-empirical mass formula, Volume 798, Issues 1–2, 1 January 2008, Pages 29–60, doi:10.1016/j.nuclphysa.2007.10.011

16. Byeongnoh Kim and Dongwoo Cha Department of Physics, Inha University, Incheon 402-751 (Received 12 April 2010) On the Mass Number Dependence of the Semi-empirical Mass Formula, Journal of the Korean Physical Society, Vol. 56, No. 5, May 2010, pp. 1546-1549

About the Authors

Ms Ankita was born 1995 in Phephana (Nohar), India, in 1995. She received her B.Sc. in 2013 and M.Sc. (Physics) in 2015 from M.G.S. University, Bikaner. She presented a paper in a national conference at MLV College, Bhilwara.

Bhuvneshwer Suthar was born in Bikaner, India, in 1983. He received the B.Sc. degree from the M.D.S. University, Ajmer, India in 2002 and the M.Sc. degree in physics in 2004 and Ph.D. degree in Physics in 2011 from the M.G.S. University (formally University of Bikaner), Bikaner, India. From 2006 to 2008, he was Junior research Fellow of CSIR in the Deptt. of Physics, Govt. Dungar College, Bikaner, India. He was Senior Research Fellow of CSIR during 2008 to 2011 in the same department. He was assistant professor of Physics, Govt. College of Engineering and Technology, Bikaner. He is currently working as Assistant Professor of Physics, Govt. Engineering College, Bikaner.

Dr. Suthar has been published more than forty research papers in various journals and conference proceedings. He has also published a book as an author and one edited book. He was also editor of three conference proceedings. His research interests include optoelectronics, photonics, and nonlinear optics etc.

Dr. Suthar is a member of International Association of Computer Science and Information Technology (IACSIT). He is also an editorial board member in two international journals and reviewer in various reputed international journals.